SURVEYING MATHEMATICS MADE SIMPLE

An original book by

Jim Crume P.L.S., M.S., CFedS

Co-Authors
Cindy Crume
Bridget Crume
Troy Ray R.L.S.
Mark Sandwick L.S.I.T.

PRINTED EDITION

*****

PUBLISHED BY:
Jim Crume P.L.S., M.S., CFedS

Coordinate Transformation

Book 9 of this Math-Series

Copyright 2013 © by Jim Crume P.L.S., M.S., CFedS

All Rights Reserved

First publication: November, 2013

Printed by CreateSpace

Available on Kindle and other devices

Cover photo courtesy of Marko Jankovic and Milorad Vasic

Geodet DB http://www.geodetdb.com/

TERMS AND CONDITIONS

The content of the pages of this book is for your general information and use only. It is subject to change without notice.

Neither we nor any third parties provide any warranty or guarantee as to the accuracy, timeliness, performance, completeness or suitability of the information and materials found or offered in this book for any particular purpose. You acknowledge that such information and materials may contain inaccuracies or errors and we expressly exclude liability for any such inaccuracies or errors to the fullest extent permitted by law.

Your use of any information or materials in this book is entirely at your own risk, for which we shall not be liable. It shall be your own responsibility to ensure that any products, services or information available in this book meet your specific requirements.

This book may not be further reproduced or circulated in any form, including email. Any reproduction or editing by any means mechanical or electronic without the explicit written permission of Jim Crume is expressly prohibited.

Table of Contents

INTRODUCTION..4

NORTHING AND EASTING TRANSFORMATION 8

ELEVATION TRANSFORMATION.........................13

SCALE TRANSFORMATION...................................17

ROTATIONAL TRANSFORMATION......................25

METHOD ONE..25

METHOD TWO...30

GRID TO GROUND TRANSFORMATION............35

NOTES...38

PRACTICAL EXAMPLE...40

SOLUTION TO PRACTICAL EXAMPLE................45

CONCLUSION..55

ABOUT THE AUTHOR...56

INTRODUCTION

Straight forward Step-by-Step instructions.

This book is just one part in a series of digital and printed editions on Surveying Mathematics Made Simple. The subject matter in this book will utilize the methods and formulas that are covered in the books that precede it. If you have not read the preceding books, you are encouraged to review a copy before proceeding forward with this book.

For a list of books in this series, please visit:

http://www.cc4w.net/ebooks.html

Prerequisites for this book:

A basic knowledge of geometry, algebra and trigonometry is required for the explanations shown in this book.

Book 1 - Bearings and Azimuths - How to add bearings and angles, subtract between bearings, convert from degrees-minutes-seconds to decimal degrees, convert from decimal degrees to degrees-minutes-seconds, convert from bearings to azimuths and convert from azimuths to bearings.

Book 2 - Create Rectangular Coordinates - How to calculate the northing and easting of an end point given the coordinates of the beginning point, bearing and distance of a line.

Book 3 - Inverse Between Rectangular Coordinates - How to determine the bearing and distance of a line given the coordinates for the beginning and ending point.

Definition:

Coordinate Transformation - [Wikipedia](): A coordinate transformation is a conversion from one datum system to another. The Cartesian coordinate system is used for surveying related datum's that consist of two and three dimensional spaces which uses two (three) numbers representing distances from an origin in two (three) mutually perpendicular directions. The two numbering system uses a Northing and Easting nomenclature. The third number element includes the Elevation for any given point.

Basic coordinate transformation models include Northing/Easting, Elevation, Scaling and Rotational or any combination of these four types. These four basic transformation models will be covered in this book. This book will also cover converting Grid to Ground coordinates, Ground to Grid coordinates and Grid Bearings to Geodetic Bearings.

Other complex coordinate transformation models such as Linear least squares, Non-linear least squares, Weighted least squares, Regression analysis and Statistics will not be covered in this book. These models require computer programs to run multiple iterations for a solution.

The Grid to Ground transformations shown in this book are those that are commonly used throughout the surveying profession and transportation agencies. These methods produce results that are within acceptable positional tolerances for surveying practices.

There are other more complex solutions for Grid to Ground transformations that address individual scale factors for each coordinate pair, forward and backward geodetic bearings along each line adjusted for convergence of the meridians. These conformal projections require computer programs to handle the more complex mathematical solutions and will not be covered in this book.

This book will focus on coordinate transformation models that most professional surveyors and engineers use on a daily basis.

A coordinate set is any group of coordinates that contain a Northing and Easting coordinate that is defined by a datum system. The datum can be autonomous, assumed or an agency defined system. The coordinate set can be measured, calculated from record information or from any other source.

The coordinate set can also be referred to as an ASCII file using surveying terms. The ASCII file is usually in a form of a Point number, Northing, Easting separated by commas with each point on a separate line. The ASCII file can also contain an Elevation and point description. The most common format is Point number, Northing, Easting,

Elevation, Description. An ASCII file is in a text format which can be opened by any text editor such as Windows Notepad and coordinate geometry programs.

NORTHING AND EASTING TRANSFORMATION

Northing and Easting transformation consist of adding a known northing datum shift value to the North coordinate and adding a known easting datum shift value to the East coordinate of the original coordinate set. No transformation is applied to the elevation component.

Figure 1 shows the transformation of an original coordinate set to a transformed coordinate set.

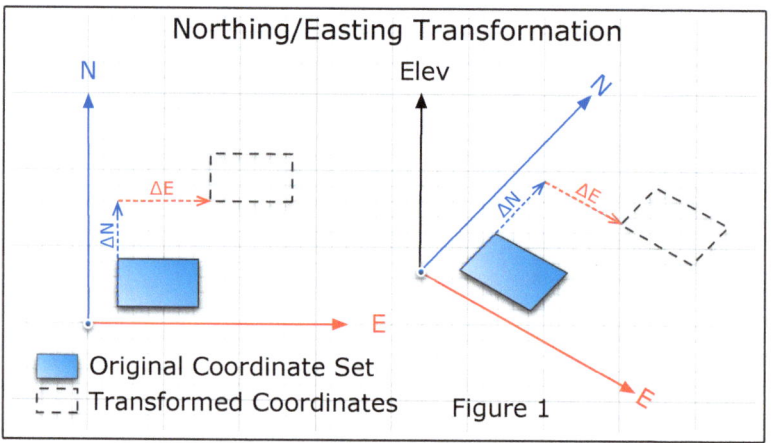

Figure 1

The purpose of a Northing/Easting transformation is to convert from one datum to another datum such as an assumed datum to an agency defined datum. The positional relationship between the coordinates in a coordinate set does not change, only the relationship to the origin of the two different datum's.

In order to convert from one datum to another datum, there must be at least one common point that is defined in both datum's such as a section corner, property corner or control point.

Formulas:

To determine the coordinate datum shift, take the differences of the Northing and Easting of a common point within both datum's.

N_o & E_o = Original Coordinate Set

N_T & E_T = Transformed Coordinates

$\Delta N = N_T - N_o$ [At the common point]

$\Delta E = E_T - E_o$ [At the common point]

$N_{T(Pt.?)} = N_{o(Pt.?)} + \Delta N$ [Repeat for each point in the coordinate set]

$E_{T(Pt.?)} = E_{o(Pt.?)} + \Delta E$ [Repeat for each point in the coordinate set]

Example:
Original Coordinate Set
Point 1 [Common Point]

N_o = 10000.00000

E_o - 20000.00000

Point 2

N_o = 10432.34214

E_o = 20563.21659

Point 3

N_o = 10325.98365

$E_0 = 20452.43259$

Transformed Coordinates

Point 1 [Common point]

$N_T = 1263421.39873$

$E_T = 598732.49343$

Solve for the remaining points in the Original Coordinate Set.

Point 2

$N_T = ???.?????$

$E_T = ???.?????$

Point 3

$N_T = ???.?????$

$E_T = ???.?????$

Determine the coordinate datum shift at the common point.

$\Delta N = 1263421.39873 - 10000.00000$

$\Delta N = \mathbf{1253421.39873}$

$\Delta E = 598732.49343 - 20000.00000$

$\Delta E = \mathbf{578732.49343}$

Apply the datum shift to the remaining points in the coordinate set.

Point 2

$N_T = 10432.34214 + 1253421.39873$

$N_T = \mathbf{1263853.74087}$

E_T = 20563.21659 + 578732.49343

E_T = **599295.71002**

Point 3

N_T = 10325.98365 + 1253421.39873

N_T = **1263747.38238**

E_T = 20452.43259 + 578732.49343

E_T = **599184.92602**

Note: Rounding error is dependent upon the number of decimal places that are utilized. It is recommended that at least 5 decimal places be used for all calculations then round the final answer as needed.

NOTES

ELEVATION TRANSFORMATION

Elevation transformation consist of adding a known elevation datum shift value to the elevation of the original coordinate set. No transformations are applied to the northing and easting component.

Figure 2 shows the transformation of an original coordinate set to a transformed coordinate set.

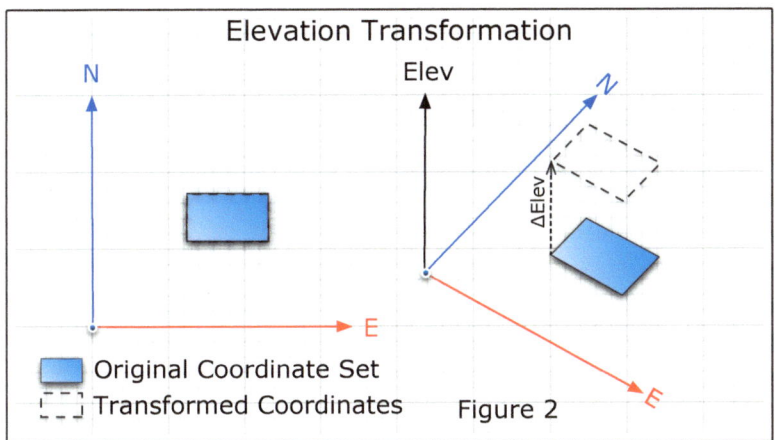

Figure 2

The purpose of an Elevation transformation is to convert from one datum to another datum such as an assumed benchmark to an agency defined benchmark.

In order to convert from one datum to another datum there must be at least one common point that is defined in both datum's such as section corner, property corner or control point.

Formulas:

To determine the elevation shift, take the differences of the elevation of the common point between both datum's.

EL_O = Original Coordinate Set

EL_T = Transformed Coordinates

$\Delta EL = EL_T - EL_O$ [At the common point]

$EL_{T(Pt.?)} = EL_{O(Pt.?)} + \Delta EL$ [Repeat for each point in the coordinate set]

Example:
Original Coordinate Set
Point 1 [Common Point]

$EL_O = 100.00$

Point 2

$EL_O = 105.52$

Point 3

$EL_O = 108.65$

Transformed Coordinates
Point 1 [Common Point]

$EL_T = 1234.98$

Solve for the remaining points in the Original Coordinate Set.

Point 2

$EL_T = ????.??$

Point 3

$EL_T = ????.??$

Determine the elevation shift at the common point.

$\Delta EL = 1234.98 - 100.00$

$\Delta EL = \mathbf{1134.98}$

Apply the datum shift to the remaining points in the coordinate set.

Point 2

$EL_T = 105.52 + 1134.98$

$EL_T = \mathbf{1240.50}$

Point 3

$EL_T = 108.65 + 1134.98$

$EL_T = \mathbf{1243.63}$

NOTES

SCALE TRANSFORMATION

Scale transformation consist of scaling the Northing and Easting of the Original Coordinate Set by a defined scale factor. There are two methods of scaling a set of coordinates. **Method One** uses the origin of (0,0) to apply the scale factor. **Method Two** uses a coordinate point within the original coordinate set as the origin.

METHOD ONE

Figure 3 shows the scale transformation of an original coordinate set to a transformed coordinate set using an origin of (0,0).

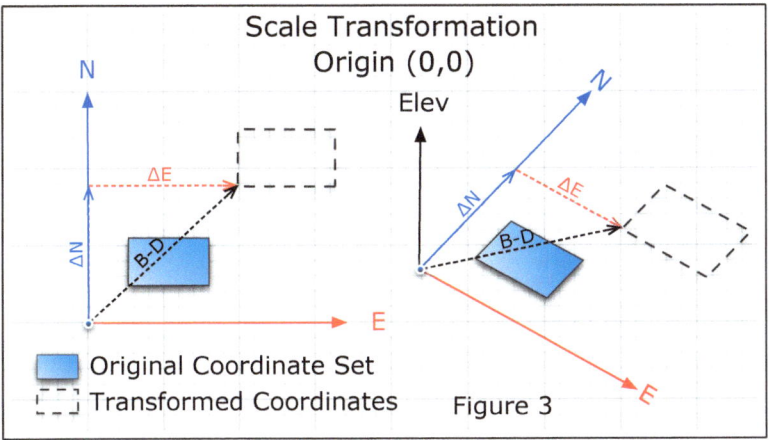

Figure 3

The purpose of a Scale transformation is to convert from one datum(unit) to another datum(unit) such as going from Chains to Feet or Feet to Meters, etc.

When working in a unit such as feet, it is sometimes easier to create coordinates using the units as shown on a BLM plat which are in chains then scale the coordinates to feet.

Formulas:

To determine the transformed coordinate multiply the Northing and Easting of the original coordinate set by the scale factor.

N_O & E_O = Original Coordinate Set

N_T & E_T = Transformed Coordinates

$N_{T(Pt.?)} = N_{O(Pt.?)}$ * Scale Factor [Repeat for each point in the coordinate set]

$E_{T(Pt.?)} = E_{O(Pt.?)}$ * Scale Factor [Repeat for each point in the coordinate set]

Example:

Scale Factor = 66.00 [Chain to Feet]

Original Coordinate Set

Point 1

N_O = 1000.00000

E_O - 2000.00000

Point 2

N_O = 1080.34214

E_O = 2001.21659

Point 3

N_O = 1080.98365

E_O = 2079.43259

Transformed Coordinates

Point 1

N_T = 1000.00000 * 66.00

N_T = **66000.00000**

E_T = 2000.00000 * 66.00

E_T = **132000.00000**

Point 2

N_T = 1080.34214 * 66.00

N_T = **71302.58124**

E_T = 2001.21659 * 66.00

E_T = **132080.29494**

Point 3

N_T = 1080.98365 * 66.00

N_T = **71344.92090**

E_T = 2079.43259 * 66.00

E_T = **137242.55094**

Note: Rounding error is dependent upon the number of decimal places that are utilized. It is recommended that at least 5 decimal places be used for all calculations then round the final answer as needed.

NOTES

METHOD TWO

Figure 4 shows the scale transformation of an original coordinate set to a transformed coordinate set using an origin of a point within the coordinate set.

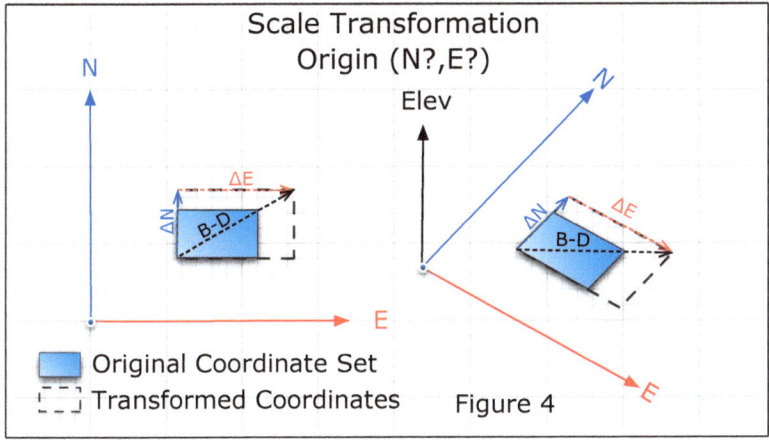

Figure 4

The purpose of an Scale transformation is to convert from one datum(unit) to another datum(unit) such as going from Chains to Feet or Feet to Meters, etc.

When working in a unit such as meters, it is sometimes easier to create coordinates using the units as shown on a legal description in feet, then scale the coordinate set to meters.

Formulas:

To determine the transformed coordinate first determine which original coordinate point to use as the origin.

N_O & E_O = Original Coordinate Set

N_T & E_T = Transformed Coordinates

$N_{T(Pt.?)} = ((N_{0(Pt.?)} - N_{ORIGIN}) * \text{Scale Factor}) + N_{T-ORIGIN}$ [Repeat for each point in the coordinate set]

$E_{T(Pt.?)} = ((E_{0(Pt.?)} - E_{ORIGIN}) * \text{Scale Factor}) + E_{T-ORIGIN}$ [Repeat for each point in the coordinate set]

Example:

Scale Factor = 0.3048 [Feet to Meter] [International Foot]

Original Coordinate Set

Point 1 [Use as origin point]

N_0 = 1000.00000 [International Feet]

E_0 = 2000.00000 [International Feet]

Point 2

N_0 = 1080.34214

E_0 = 2001.21659

Point 3

N_0 = 1080.98365

E_0 = 2079.43259

Transformed Coordinates

Point 1 [Origin point]

N_T = 3000.00000 [Meter]

E_0 - 4000.00000 [Meter]

Point 2

N_T = ((1080.34214 - 1000.00000) * 0.3048) + 3000.00000

N_T = **3024.48828**

$E_T = ((2001.21659 - 2000.00000) * 0.3048) + 4000.00000$

$E_T = \mathbf{4000.37082}$

Point 3

$N_T = ((1080.98365 - 1000.00000) * 0.3048) + 3000.00000$

$N_T = \mathbf{3024.68382}$

$E_T = ((2079.43259 - 2000.00000) * 0.3048) + 4000.0000$

$E_T = \mathbf{4024.21105}$

Note: Rounding error is dependent upon the number of decimal places that are utilized. It is recommended that at least 5 decimal places be used for all calculations then round the final answer as needed.

NOTES

ROTATIONAL TRANSFORMATION

Rotational transformation consist of rotating the Northing and Easting of the Original Coordinate Set by a defined angular value. There are two methods of rotating a set of coordinates. Method one uses the origin of (0,0) to apply the angular value. Method two uses a coordinate point within the original coordinate set as the origin.

METHOD ONE

Figure 5 shows the rotational transformation of an original coordinate set to a transformed coordinate set using an origin of (0,0).

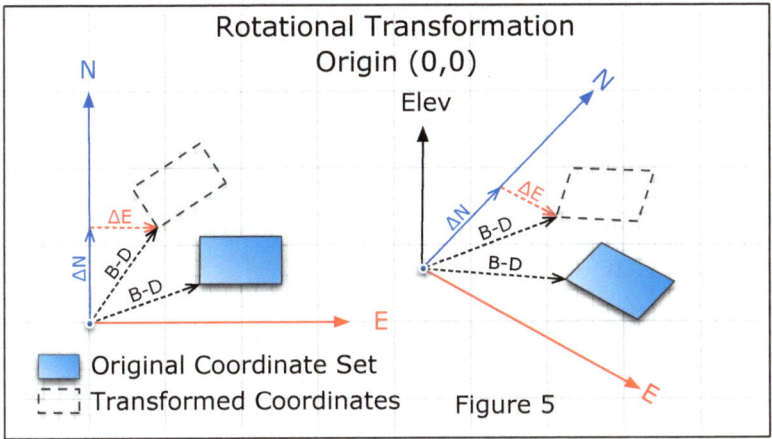

The purpose of a rotational transformation is to convert from one datum to another datum such as going from grid bearing to geodetic or true bearing.

When surveying with a GPS you will most likely be measuring in grid bearings. Without going into the details of multiple point calibration, let's assume that you are working on state plane coordinates with grid bearings.

Formulas:

To determine the transformed coordinate, rotate the Northing and Easting of the original coordinate set by the angular value.

N_O & E_O = Original Coordinate Set

N_T & E_T = Transformed Coordinates

$Distance_{(O)(Course?)} = \sqrt{N_{O(Pt.?)}^2 + E_{O(Pt.?)}^2}$

$Bearing_{(O)(Course?)} = ArcTan(E_{O(Pt.?)} / N_{O(Pt.?)})$

$Bearing_{(T)(Course?)} = Bearing_{(O)(Course?)} +/- $ Angular Value

$N_{T(Pt.?)} = Cos(Bearing_{(T)(Course?)}) * Distance_{(O)(Course?)}$
[Repeat for each point in the coordinate set]

$E_{T(Pt.?)} = Sin(Bearing_{(T)(Course?)}) * Distance_{(O)(Course?)}$
[Repeat for each point in the coordinate set]

Example:

Angular value = -0°23'12"

Original Coordinate Set

Point 1

N_O = 1000.00000

E_O - 2000.00000

Point 2

N_O = 1080.34214

E_O = 2001.21659

Point 3

$N_0 = 1080.98365$

$E_0 = 2079.43259$

Transformed Coordinates
Point 1

Distance$_{(O)}$ = √(1000.00000² + 2000.00000²)

Distance$_{(O)}$ = **2236.06798**

Bearing$_{(O)}$ = ArcTan(2000.00000 / 1000.00000)

Bearing$_{(O)}$ = **63.43495° or 63°26'05.8"**

Bearing$_{(T)}$ = 63°26'05.8" - 0°23'12"

Bearing$_{(T)}$ = **63.04828° or 63°02'53.8"**

N_T = Cos(63°02'53.8") * 2236.06798

N_T = **1013.47449**

E_T = Sin(63°02'53.8") * 2236.06798

E_T = **1993.20583**

Point 2

Distance$_{(O)}$ = √(1080.34214² + 2001.21659²)

Distance$_{(O)}$ = **2274.20469**

Bearing$_{(O)}$ = ArcTan(2001.21659 / 1080.34214)

Bearing$_{(O)}$ = **61.63793° or 61°38'16.6"**

Bearing$_{(T)}$ = 61°38'16.6" - 0°23'12"

Bearing$_{(T)}$ = **61.25128° or 61°15'04.6"**

N_T = Cos(61°15'04.6") * 2274.20469

N_T = **1093.82245**

E_T = Sin(61°15'04.6") * 2274.20469

E_T = **1993.88049**

Point 3

Distance$_{(O)}$ = $\sqrt{1080.98365^2 + 2079.43259^2}$

Distance$_{(O)}$ = **2343.62231**

Bearing$_{(O)}$ = ArcTan(2079.43259 / 1080.98365)

Bearing$_{(O)}$ = **62.53254° or 62°31'57.1"**

Bearing$_{(T)}$ = 62°31'57.1" - 0°23'12"

Bearing$_{(T)}$ = **62.14587° or 62°08'45.1"**

N_T = Cos(62°08'45.1") * 2343.62231

N_T = **1094.99255**

E_T = Sin(62°08'45.1") * 2343.62231

E_T = **2072.08997**

All angles must be converted to Decimal Degrees prior to performing trigonometric operations. See Book 1 - "Bearings and Azimuths" for methods on converting Degrees-Minutes-Seconds to Decimal Degrees and vice versa. Also see Book 1 for adding and subtracting bearings and angles.

NOTES

METHOD TWO

Figure 6 shows the rotational transformation of an original coordinate set to a transformed coordinate set using an origin of a point within the coordinate set.

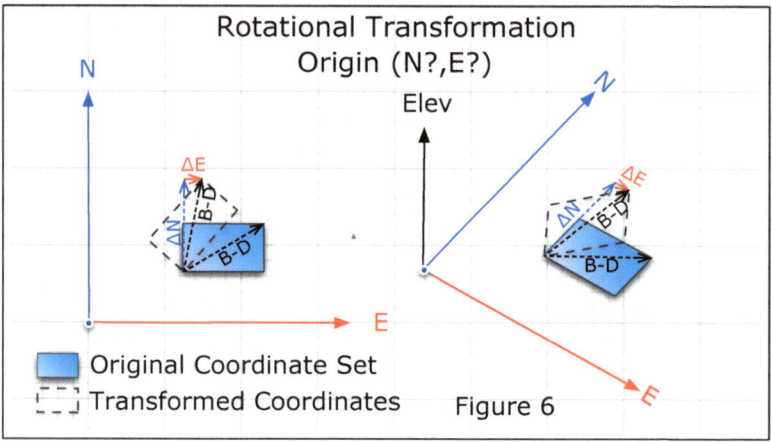

Figure 6

The purpose of a rotational transformation is to convert from one datum to another datum such as going from one basis of bearing to another basis of bearing.

When two contiguous legal descriptions have different basis of bearings along the same controlling line (such as a section line), it will be required to rotate the set of coordinates for one of the legal descriptions and transform them to the same basis of bearing of the adjacent legal description.

Formulas:

To determine the transformed coordinate, first determine which original coordinate point to use as the origin.

N_O & E_O = Original Coordinate Set

N_T & E_T = Transformed Coordinates

$\text{Distance}_{(O)(Course?)} = \sqrt{(N_{O(Pt.?)} - N_{ORIGIN})^2 + (E_{O(Pt.?)} - E_{ORIGIN})^2}$

$\text{Bearing}_{(O)(Course?)} = \text{ArcTan}((E_{O(Pt.?)} - E_{ORIGIN}) / (N_{O(Pt.?)} - N_{ORIGIN}))$

$\text{Bearing}_{(T)(Course?)} = \text{Bearing}_{(O)(Course?)} +/- \text{Angular Value}$

$N_{T(Pt.?)} = N_{ORIGIN} + (\text{Cos}(\text{Bearing}_{(T)(Course?)}) * \text{Distance}_{(O)(Course?)})$ [Repeat for each point in the coordinate set]

$E_{T(Pt.?)} = E_{ORIGIN} + (\text{Sin}(\text{Bearing}_{(T)(Course?)}) * \text{Distance}_{(O)(Course?)})$ [Repeat for each point in the coordinate set]

Example:

Angular value = -0°23'12"

Original Coordinate Set

Point 1 [Origin point]

N_O = 1000.00000

E_O = 2000.00000

Point 2

N_O = 1080.34214

E_O = 2001.21659

Point 3
N_0 = 1080.98365

E_0 = 2079.43259

Transformed Coordinates
Pt. 1 [Origin point]

N_0 = 1000.00000

E_0 = 2000.00000

Point 2
Distance$_{(O)}$ = √((1080.34214 - 1000.00000)² + (2001.21659 - 2000.00000)²)

Distance$_{(O)}$ = **80.35135**

Bearing$_{(O)}$ = ArcTan((2001.21659 - 2000.00000) / (1080.34214 - 1000.00000))

Bearing$_{(O)}$ = **0.86754° or 0°52'03.1"**

Bearing$_{(T)}$ = 0°52'03.1" - 0°23'12"

Bearing$_{(T)}$ = **0.48086° or 0°28'51.1"**

N_T = 1000.00000 + (Cos(0°28'51.1") * 80.35135)

N_T = **1080.34852**

E_T = 2000.00000 + (Sin(0°28'51.1") * 80.35135)

E_T = **2000.67435**

Point 3
Distance$_{(O)}$ = √((1080.98365 - 1000.00000)² + (2079.43259 - 2000.00000)²)

Distance$_{(O)}$ = **113.43671**

Bearing$_{(O)}$ = ArcTan((2079.43259 - 2000.00000) / (1080.98365 - 1000.00000))

Bearing$_{(O)}$ = **44.44603° or 44°26'45.7"**

Bearing$_{(T)}$ = 44°26'45.7" - 0°23'12"

Bearing$_{(T)}$ = **44.05936° or 44°03'33.7"**

N$_T$ = 1000.00000 + (Cos(44°03'33.7") * 113.43671)

N$_T$ = **1081.51786**

E$_T$ = 2000.00000 + (Sin(44°03'33.7") * 113.43671)

E$_T$ = **2078.88426**

All angles must be converted to Decimal Degrees prior to performing trigonometric operations. See Book 1 - "Bearings and Azimuths" for methods on converting Degrees-Minutes-Seconds to Decimal Degrees and vice versa. Also see Book 1 for adding and subtracting bearings and angles.

NOTES

GRID TO GROUND TRANSFORMATION

Grid to Ground transformation consist of scaling the Northing and Easting of the Grid Coordinates by a defined combined scale factor.

Details of the combined scale factor will not be covered in this book. It is assumed that a combined scale factor has already been determined for the local projection that you are working in.

Method One and **Two** as described in the **Scale Transformation** shown earlier in this book will be applied. **Method One** is most commonly used with an origin of (0,0).

Figure 7 shows the scale transformation of grid coordinates to ground coordinates using an origin of (0,0).

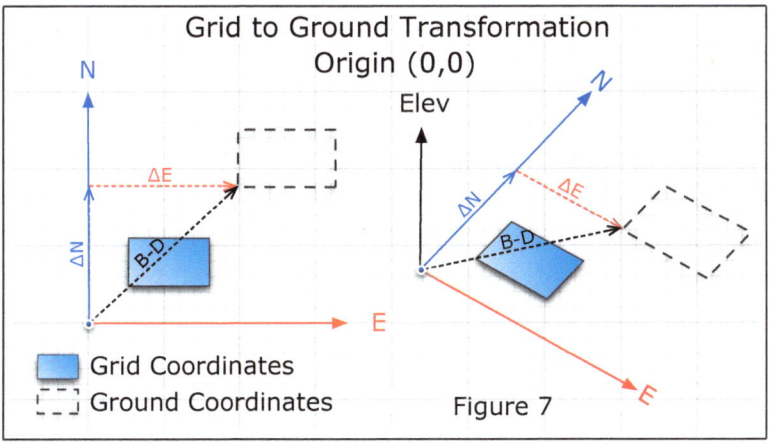

Figure 7

The scaling process and formulas are the same as **Scale Transformation** that was described earlier in this book. The only difference is the scale factor. The combined scale factor will be determined by the location of the grid coordinates within a state plane zone.

The intricate details of state plane coordinates will not be discussed in this book. A new book is being worked on titled "Understanding State Plane Coordinate System" that will described more details related to this system. Look for this book at Amazon.com

The combined scale factor can be derived by utilizing GPS software such as Trimble Business Center, as shown on an NGS Data Sheets or as defined by a private/public agencies.

To distinguish between the combined scale factor and its inverse, Grid Adjustment Factor (GAF) for the inversed value will be utilized in this book.

For most locations within a state plane zone, the GAF will be over 1.00000 meaning that the ground coordinates will be larger than the grid coordinates after the transformation. Another way to look at it is that the grid plane is below the surface for most of the state plane zone. Out on the fringes of the zone, the grid plane will be above the surface at lower elevations.

The combined scale factor shown on the NGS Data Sheets is from Ground to Grid using the form Grid

Coordinate = Ground Coordinate * Combined Scale Factor.

The inverse of the combined scale factor (GAF) is needed for the example below.

Example:

Combined Scale Factor from NGS Data Sheet = 0.99985706

Note: Eight (8) decimal places are sufficient enough for most surveying projects.

GAF = 1 / 0.99985706

GAF = 1.00014296

Ground Coordinate = Grid Coordinate * GAF

Note: It is very important that the GAF be applied in the correct direction. It is helpful to think of the ground distances will be longer than grid distances for most locations within the state plane zone.

For survey projects located out on the fringes of the state plane zone the GAF might be less than 1.00000 depending on the elevation of the project. When you see a combined scale factor over 1.00000 on an NGS Data Sheet that means that the grid plane is above the surface and the GAF after inversing the combined scale factor will be less than 1.00000. In this scenario the ground distances will be shorter than the grid distances.

Figure 8 shows the relationship of the Grid Plane to the Ground Surface.

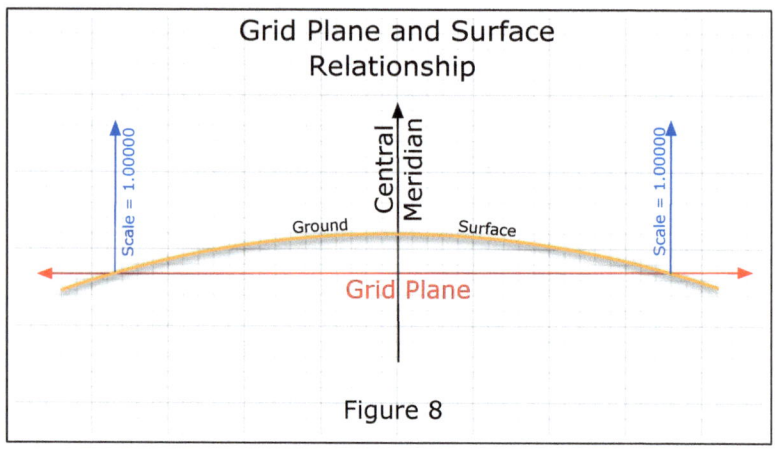

Figure 8

NOTES

GROUND TO GRID TRANSFORMATION

Ground to Grid transformation consist of scaling the Northing and Easting of the Ground Coordinates by a defined combined scale factor.

Method One and **Two** as described in the **Scale Transformation** shown earlier in this book can be applied. **Method One** is most commonly used with an origin of (0,0).

Figure 9 shows the scale transformation of ground coordinates to grid coordinates using an origin of (0,0).

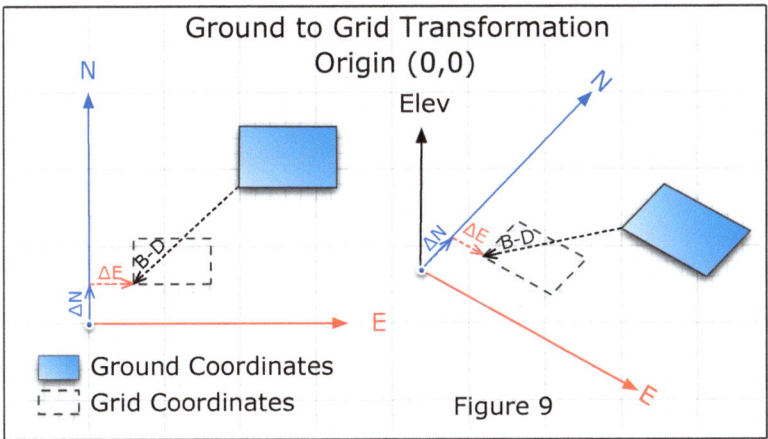

Refer to **Grid to Ground Transformation** described earlier in this book for definitions and explanations on state plane coordinates, combined scale factor and Grid Adjustment Factor (GAF).

Example:

Combined Scale Factor from NGS Data Sheet = 0.99985706

Note: Eight (8) decimal places are enough for most survey projects.

Grid Coordinate = Ground Coordinate * combined scale factor or

Grid Coordinate = Ground Coordinate / GAF

Note: It is very important that the combined scale factor be applied in the correct direction. It is helpful to think of the grid distances will be shorter than ground distances for most locations within the state plane zone.

PRACTICAL EXAMPLE

Your client has assigned you a Scope of Work to survey a of portion of a section as follows:

SCOPE OF WORK

Section 36, T.5N., R.2E.

1. Locate the West one quarter corner
2. Locate the Northwest section corner
3. Locate the North one quarter corner

The Northwest section corner will be held as the primary horizontal and vertical control using the following Grid Coordinates and Elevation:

NAD 1983 - Arizona Central Zone

N = 996909.346

E = 639075.225

Elev = 1499.398 [NAVD88]

Convergence Angle = -0°06'41"

Grid Adjustment Factor = 1.00016

DELIVERABLES

Modified Ground Coordinates by subtracting 550000.000 from the North Ground Coordinates and rotating to Geodetic North at the primary control point.

Point No. 1 (Northwest section corner)

N = ???

E = ???

Elev = ???

Point No. 2 (West one quarter corner)

N = ???

E = ???

Elev = ???

Point No. 3 (East one quarter corner)

N = ???

E = ???

Elev = ???

The sectional monumentation associated with this scope of work is located in a busy highway therefore you will be unable to set your GPS base on the primary control point (NW Sec. Cor.). You set your GPS base near the project sight and start it using an autonomous position using the NAD 1983 - Arizona

Central Zone. You collect coordinate information for the sectional monumentation based upon an autonomous position at the GPS base as follows:

Point No. 1

N = 996924.81198

E = 639080.54987

Elev = 1495.033

Point No. 2

N = 994282.72198

E = 639103.47987

Elev = 1476.526

Point No. 3

N = 996944.31398

E = 641722.99387

Elev = 1517.902

The following are the steps required to solve for the Deliverables:

Step 1

Adjust the autonomous coordinates to grid coordinates holding the primary control point (NW Sec. Cor.).

Step 2

Adjust the elevations holding the primary control point.

Step 3

Adjust the grid coordinates to ground coordinates using the GAF.

Step 4

Adjust the north ground coordinate to the modified ground coordinate for each point by subtracting 550000.000 from the northing.

Step 5

Rotate the modified ground coordinates using the primary control point as the rotation point using the convergence angle.

The solution can be found near the end of this book.

NOTES

SOLUTION TO PRACTICAL EXAMPLE

Step 1 - Autonomous to Grid Transformation

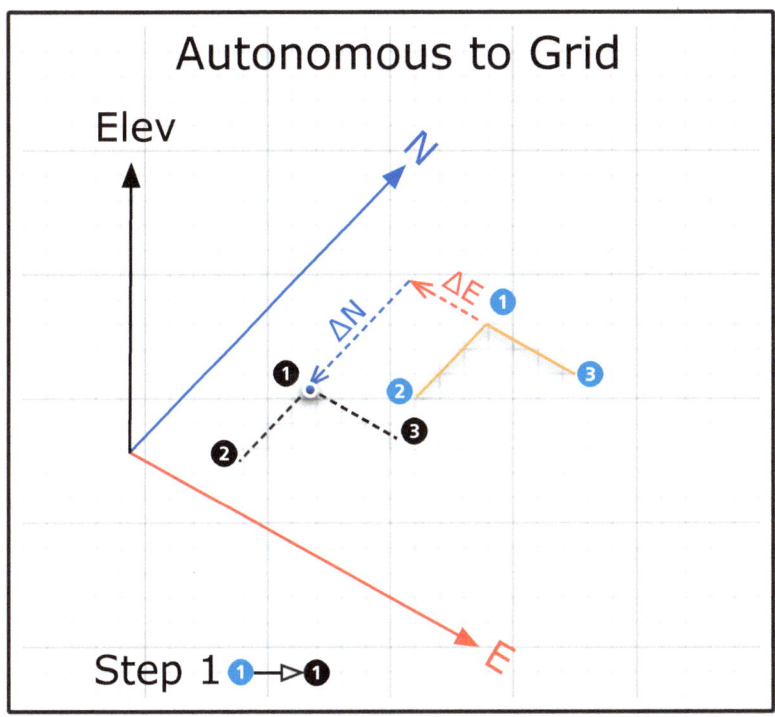

The first step is to adjust the autonomous positions for the corners to grid coordinates by determining the amount of datum shift in the Northerly and Easterly direction from position No. 1 to the Northwest section corner (Control point).

Determine the coordinate datum shift at the common point.

Datum Shift

ΔN = 996909.346 - 996924.81198

ΔN = **-15.46598**

ΔE = 639075.225 - 639080.54987

ΔE = **-5.32487**

Point No. 1

N = **996909.34600**

E = **639075.22500**

Point No. 2

N = 994282.72198 - 15.46598

N = **994267.25600**

E = 639103.47987 - 5.32487

E = **639098.15500**

Point No. 3

N = 996944.31398 - 15.46598

N = **996928.84800**

E = 641722.99387 - 5.32487

E = **641717.66900**

Step 2 - Elevation Transformation

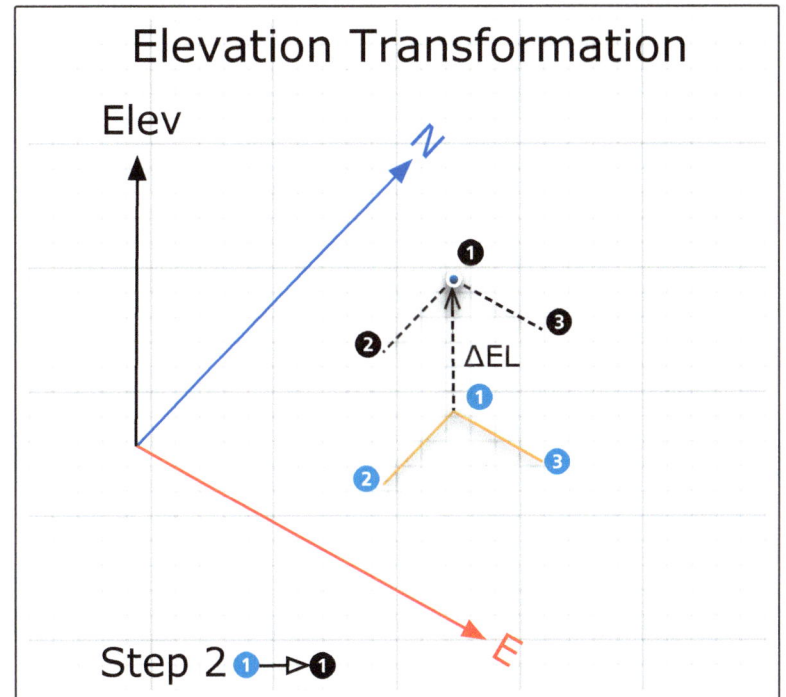

Determine the elevation shift at the common point.

Elevation Shift

ΔEL = 1499.398 - 1495.033

ΔEL = **4.365**

Point No. 1

Elev = **1499.398**

Point No. 2

Elev = 1476.526 + 4.365

Elev = **1480.891**

Point No. 3

Elev = 1517.902 + 4.365

Elev = **1522.267**

Step 3 - Grid to Ground Transformation

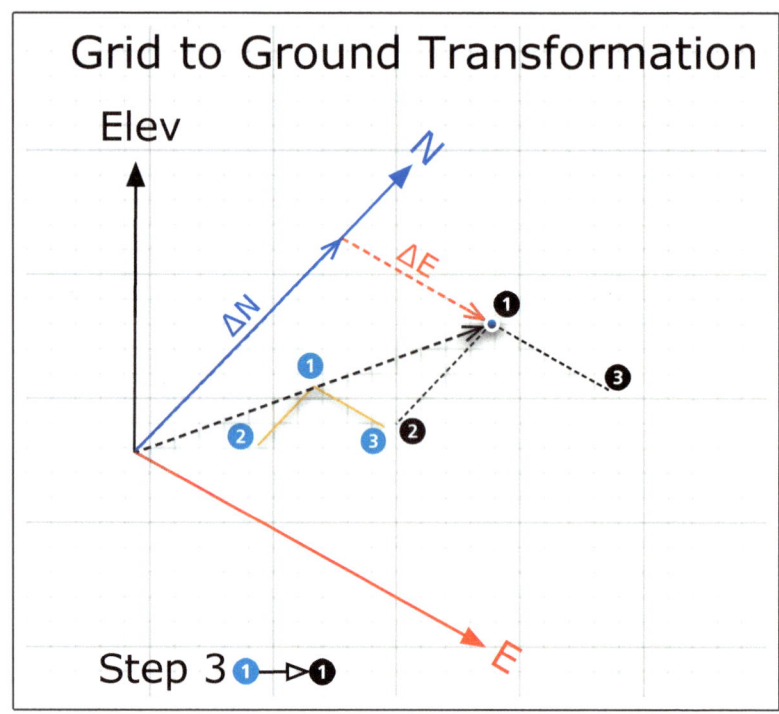

Determine the ground coordinates by multiply the Northing and Easting of the grid coordinates by the GAF.

Ground Coordinates

Point No. 1

N = 996909.34600 * 1.00016

N = **997068.85150**

E = 639075.22500 * 1.00016

E = **639177.47704**

Point No. 2

N = 994267.25600 * 1.00016

N = **994426.33876**

E = 639098.15500 * 1.00016

E = **639200.41070**

Point No. 3

N = 996928.84800 * 1.00016

N = **997088.35662**

E = 641717.66900 * 1.00016

E = **641820.34383**

Step 4 - Modified Ground Coordinates

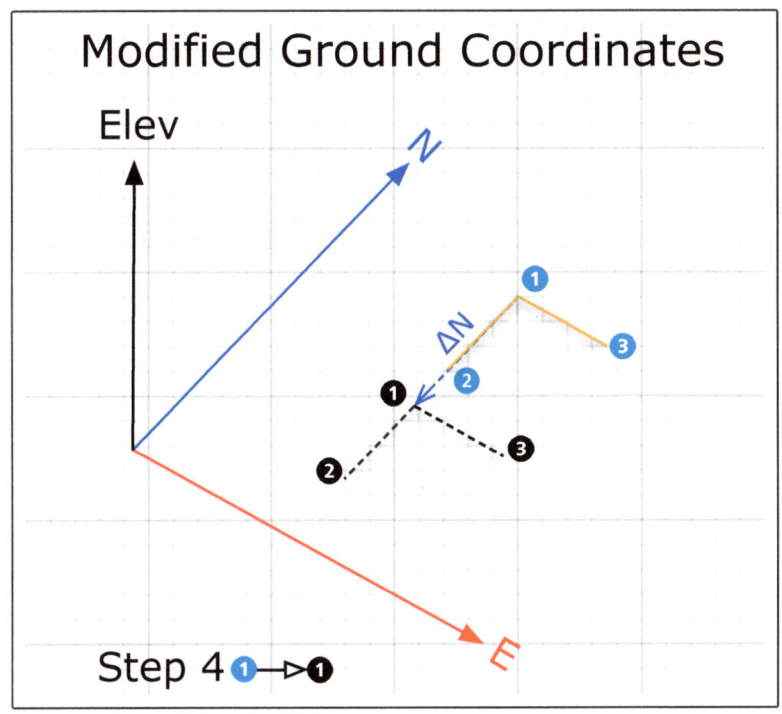

Subtract 550000.000 from the North Ground Coordinate from each point.

Point No. 1

N = 997068.85150 - 550000.000

N = **447068.85150**

E = **639177.47704**

Point No. 2

N = 994426.33876 - 550000.000

N = **444426.33876**

E = **639200.41070**

Point No. 3

N = 997088.35662 - 550000.000

N = **447088.35662**

E = **641820.34383**

Step 5 - Grid Bearings to Geodetic Bearings Transformation

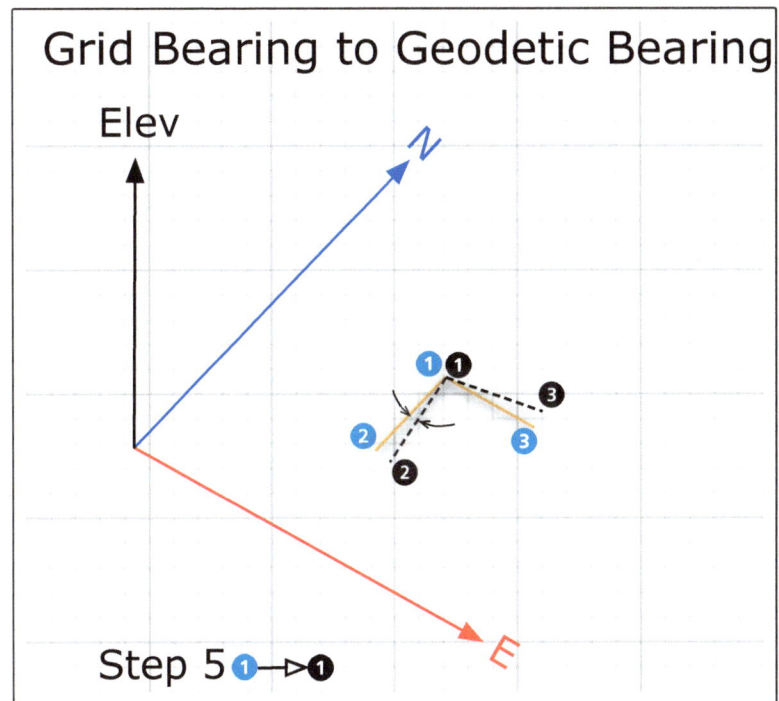

Holding Point 1 as the Origin, determine the Bearing and Distance from the origin to Points 2 and 3. Then rotate the Bearing by the convergence angle. Calculate the Latitude and Departure using the Geodetic Bearing and Distance then add to the Northing and Easting of the Origin Point.

Point 1 [Origin point]

N = 447068.85150

E = 639177.47704

Point 2

Distance = $\sqrt{((444426.33876 - 447068.85150)^2 + (639200.41070 - 639177.47704)^2)}$

Distance = **2642.61226**

Bearing = ArcTan((639200.41070 - 639177.47704) / (444426.33876 - 447068.85150))

Bearing = **S0.49724E° or S00°29'50.0"E**

Bearing = S0°29'50.0E" - (-0°06'41")

Bearing = **S0.60861E° or S00°36'31.0E"**

N = 447068.85150 - (Cos(0°36'31.0") * 2642.61226)

N = **444426.38832**

E = 639177.47704 + (Sin(0°36'31.0") * 2642.61226)

E = **639205.54705**

Point 3

Distance = √((447088.35662- 447068.85150)² + (641820.34383- 639177.47704)²)

Distance = **2642.93877**

Bearing = ArcTan((641820.34383- 639177.47704) / (447088.35662- 447068.85150))

Bearing = **N89.57715E° or N89°34'37.7"E**

Bearing = N89°34'37.7"E + (-0°06'41")

Bearing = **N89.46575E° or N89°27'56.7"E**

N = 447068.85150 + (Cos(89°27'56.7") * 2642.93877)

N = **447093.49502**

E = 639177.47704 + (Sin(89°27'56.7") * 2642.93877)

E = **641820.30092**

DELIVERABLES

Point No. 1 (Northwest section corner)

N = 447068.85150

E = 639177.47704

Elev = 1499.398

Point No. 2 (West one quarter corner)

N = 444426.38832

E = 639205.54705

Elev = 1480.891

Point No. 3 (East one quarter corner)

N = 447093.49502

E = 641820.30092

Elev = 1522.267

Note: Rounding error is dependent upon the number of decimal places that are utilized. It is recommended that at least 5 decimal places be used for all calculations then round the final answer as needed.

All angles must be converted to Decimal Degrees prior to performing trigonometric operations. See Book 1 - "Bearings and Azimuths" for methods on converting Degrees-Minutes-Seconds to Decimal Degrees and vice versa. Also see Book 1 for adding and subtracting bearings and angles.

NOTES

CONCLUSION

Coordinate Transformation is a vital part of everyday surveying practice. Having a thorough understanding of the methods for the basic transformation models is critical to being a Professional Land Surveyor. The book has laid out the basic transformation models that you will encounter during your surveying career. You will utilize computer software to perform these models for the most part. Having the knowledge of how it all works will help you to appreciate the mathematics behind the solutions.

ABOUT THE AUTHOR
Jim Crume P.L.S., M.S., CFedS

My land surveying career began several decades ago while attending Albuquerque Technical Vocational Institute in New Mexico and has traversed many states such as Alaska, Arizona, Utah and Wyoming. I am a Professional Land Surveyor in Arizona, Utah and Wyoming. I am an appointed United States Mineral Surveyor and a Bureau of Land Management (BLM) Certified Federal Surveyor. I have many years of computer programming experience related to surveying.

This book is dedicated to the many individuals that have helped shape my career. Especially my wife Cindy. She has been my biggest supporter. She has been my instrument person, accountant, advisor and my best friend. Without her, I would not be the professional I am today. Cindy, thank you very much.

Other titles by this author:

http://www.cc4w.net/ebooks.html

www.ingramcontent.com/pod-product-compliance
Lightning Source LLC
Chambersburg PA
CBHW040848180526
45159CB00001B/358